The Origin of Galaxies

Jin He

authorHOUSE®

AuthorHouse™
1663 Liberty Drive
Bloomington, IN 47403
www.authorhouse.com
Phone: 1-800-839-8640

First published by AuthorHouse 11/21/2009

ISBN: 978-1-4490-4118-2 (sc)

Printed in the United States of America
Bloomington, Indiana

This book is printed on acid-free paper.

The Origin of Galaxies

– Origin Exploration 4 –

by Jin He

© Jin He , October 8th 2009

The Origin of Galaxies

– Origin Exploration 4 –

by Jin He

Contents

i

List of Figures

Chapter 1

Human Crisis and the Origin of Cosmic Dust

1.1   Human Crisis on the Earth

The history of human on the Earth is at least several hundred million years. However, we do not know how human is originated. Human beings seem to be the wandering street children who have lost their parents. The most horrible thing is that human beings never realize that they are the wandering street children. We know neither our blood lineage nor the meaning of life. We do not know how to value our own or other's lives. Such blind and cruel actions as deceit, plunder, warfare, and so on, have strangled millions and generations of lives.

Today, the global world needs a reflection on itself and mankind needs to resort to its starting point. It is time for human beings to seek their origin and answer such primary questions as what human body is made of and what humanity is. Further more, the crises that the Earth's inhabitants encounter require mankind's answer to these questions!

Towards the 21st century, mankind on earth encounters unprecedented crises. The natural environment suffers the most important crisis which may reclaim mankind's fate. The crisis manifests itself with air and river pollution, ecological imbal-

3

ance, as well as the global warming caused by greenhouse effect. The second crisis is the financial and social crisis caused by human's bubble economy. The resolution of these crises turns out to be mutual contradiction. In addition to these crises, the ever-changing material life has also brought the devastating effect on human body and mind.

Reviewing through human history, we can see that the crises stem from human's ignorance of their own origin. If human beings know the origin, then all kinds of crises can be prevented, the existing crises can be mitigated and even eliminated.

Very fortunately, human understanding of the nature over the last centuries is enough for human beings to understand the origin. Physics, chemistry, biology and medicine have fully proved that human bodies are composed of elementary particles. The question is, what force combines elementary particles into the biosphere?

1.2 The Power which Governs Everything: Gravitation

Scientists have fully proved that there exist only four forces among particles: electromagnetic, weak nuclear, strong nuclear, and gravitational. The nuclear forces are short-ranged while the electromagnetic force is long-ranged. Each of the three has two contradictory aspects of attracting and rebeling and, generally, has no net effect in the macro-world due to offsetting effect. By artificially destroying the offsetting effect, scientists and engineers can make scientific or commercial products based on the earth's natural structures. Atomic bombs result from artificially destroying the offsetting effect of nuclear force. Computers, telephones, TV sets and so on are the examples of artificially destroying the offsetting effect of electromagnetic force.

Gravitational force, however, has no contradictory aspects. Gravity has the only effect of attraction and, therefore, can not offset itself. Because of this, the true origin of natural structure is the gravitational force.

The origin of natural structure can not be other forces. Modern science has fully proved that independent system of microscopic particles combined by electromagnetic force or nuclear forces inevitably moves towards chaotic state rather than orderly one. This is the principle of entropy increase, which is well known for scientists. Therefore, if there were no gravitational force then the whole universe would be simply uniform gas without structure. However, there exist in the macroscopic world such orderly structures large as galaxies and small as stars, planets, plants, animals, and even human beings. Therefore, varied kinds of macro-world structures result from the struggling of the gravitational force against the electromagnetic and nuclear forces.

Unfortunately, gravity is very very weak. For example, it is $0.0000 \cdots 00001$ times (where 40 zeros are after the decimal point) weaker than the electricity! Only the Earth, Moon, Sun and so on present gravity. There is no slightest gravity between cars or human bodies. Therefore, human beings suffer insurmountable difficulty to experimentally study gravity. Because human beings are insignificant, it is impossible to do physical experiments on such macroscopic systems as the solar system or galaxies. Even if human beings could do such experiment, they would not have sufficient time to complete it. We should know that the life of the Sun is about billion years!

However, we can use man-made telescopes to take images of large-scale material systems such as galaxies. We can analyze the images!

1.3 The Material which Constructs Human body: Dust

Human beings can understand the gravity by studying the large-scale structures in the universe. Both human bodies and planets are mainly composed of the elements which are heavier than hydrogen and helium. On the contrary, the Sun is mainly composed of the lightest elements, i.e., hydrogen and helium. Therefore, human beings could neither live on the Sun nor be originated from it. However, cosmic dust is similar to human bodies in their constituents, and is mainly composed of the heavier elements. Therefore, any planet in the universe must be derived from cosmic dust. Of course, humans beings must be derived from cosmic dust. This confirms what Bible says: man is made of earth! Therefore, if we know the origin of cosmic dust, we must know the origin of humanity!

Very fortunately, it is very easy for us to see where is dust or no dust by looking into the image of any physical system in the universe. The basic constituents in the universe are galaxies. Therefore, knowing how dust is generated in galaxies is equivalent to knowing how lives are originated. And we can even know more.

Do not be fooled with color images. Some people are indulged in women's pretty looks, but they simply do not know what is color. Color is essentially the different frequencies or wavelengths of light. In fact, the shape of an object or its photo is the distribution of light arriving at your eyes from the surface of the object. That is, it is the distribution of light frequency and density varying with the surface of the object. Light of longer wavelength that appears reddish has strong penetrating ability. In other words, reddish light refuses to be absorbed by dust or gas. Elliptical galaxies are very clean, with no observation of gas and dust. Therefore, it does not matter to catch which color for you to take the images of elliptical galaxies. Images of the same elliptical galaxy of different colors are very similar and smooth. They are the good demonstration of star distribution in the galaxy. But elliptical galaxies are three-dimensional while their images are two-dimensional. The image of an elliptical galaxy is the cumulative density of stars in the observing directions.

Spiral galaxies are just the opposite. They have a large amount of gas and dust. Although their shapes are two-dimensional, they have a certain degree of thickness. Therefore, if we take images of spiral galaxies in the shorter wavelength (i.e., bluish light) then the light from the stars that are behind gas and dust are basically absorbed by the gas and dust. As a result, the image is mainly the distribution of gas and dust. Because the distribution of gas and dust is not smooth, the image looks ugly. Internet images of spiral galaxies are usually short-wavelength ones, therefore, people are daunted by the mysterious look of gas and dust (see the lower-panel of Fig. 1.1). Therefore, to get an image of spiral galaxy which is mainly stellar density distribution, we take light of longer wavelength from the galaxy, e.g., infrared image. The resulting image is reddish. Although gas and dust have charming and bright colors, they have negligible mass.

Figure 1.1: Upper panel: the infrared image of normal spiral galaxy M51 (image credit [1]). Lower panel: the blue-band image of the same galaxy (image credit [2]).

1.4 The Origin of Galaxies: Rational Gravity

The usually familiar gravity refers to the force which exerts between Earth and Moon or between Sun and Earth. These examples of gravity are the interaction between two bodies. As for the behavior of gravity exerted on many bodies, the solar system can not be the example for us to study such behavior. However, each galaxy is composed of billions of stars, which demonstrates the gravitational interaction among many bodies. Galaxy images show that each galaxy has a center. Star density at the center is the highest. From the center outward, the density is smaller and smaller and presents a regular pattern, known as galaxy structure. The principle behind the formation of galaxy structures is the demonstration of gravitational interaction in many-body systems.

Then, what is the behavior of gravitational interaction in many-body systems? Dr. He pioneered the study on galaxy structures in 2003 and the study shows that stars in any galaxy are controlled by a very simple orderly force involving many-bodies: proportion. Because solar system is just a point at the Milky Way galaxy, the proportion force reduces to Newtonian gravity between two bodies!

Proportion means that the distribution of matters in the universe is orderly. For example, there are four giants standing in array. Their heights are respectively A, B, C, D, and A, B stand in the first row from left to right, C, D in the second row from left to right. According to the view of mainstream cosmologists, the four giants can have any heights and can stand at any position. That is why current foundational scientific theories are incompetent and can not explain the origin of natural structures! They can not provide any basic principle to explain such orderly structure as human beings nor to resolve the motion of the most simple gravitational systems (such as interactional free three-bodies). However, the universe is orderly. The orderly force at the largest scales requires that the distribution of heights is in proportion. In other words, A divided by B is equal to C divided by D. This means that A divided by C is equal to B divided by D. If there are nine giants standing in array, then the ratios of heights from neighboring two rows are constant (proportion rows). Similarly, the ratios of heights from neighboring two lines are constant (proportion lines). In this way are galaxies created!

The above-said rows and lines are all straight (proportion lines). but each galaxy is a regional distribution of matters in the universe. Therefore, the proportion lines of each galaxy are curved but the rows and columns still cross at each other vertically and they form the net of orthogonal curves.

In general, a distribution of similar bodies is called the rational structure if its density varies proportionally along some particular net of orthogonal curves. In book 1 of Origin Exploration [3], rational structure is called the matriarchal structure. From now on we call it rational structure. Therefore, independent galaxies are all rational structures. The force which leads to the rational structure is called the rational force, i.e., the proportion force which is the demonstration of gravity in large-scale and many-body system. That is, the universe is rational. This is the most important discovery in human history.

Chapter 2

Two Examples of Rational Structure

2.1 Logarithmic Density of Galaxy Structure

Now we start the scientific and mathematical investigation into galaxy structures. A galaxy is a distribution of stars. But we can not see individual stars on a galaxy image. A galaxy image is the distribution of star densities. Therefore, we use a mathematical function to describe a distribution of densities. Because spiral galaxies are planar, we use a function of two variables, x, y, to describe the stellar distribution of a spiral galaxy:

$$\rho(x, y) \tag{2.1}$$

where x, y is the rectangular Cartesian coordinates on the spiral galaxy plane. The coordinate origin is the galaxy center. Therefore, $\rho(0, 0)$ is the stellar density at the galaxy center. We want to study the ratio of the density ρ_2 to the density ρ_1 at two positions 2 and 1 respectively:

$$\rho_2/\rho_1. \tag{2.2}$$

In fact, the logarithm of the ratio divided by the distance s between the two positions is approximately the directional derivative of the logarithmic density $(f(x,y) = \ln \rho(x,y))$ along the direction of the two positions:

$$(\ln(\rho_2/\rho_1))/s = (\ln \rho_2 - \ln \rho_1)/s \approx \frac{\partial f}{\partial s} \qquad (2.3)$$

There is no systematic mathematical theory on ratios. Therefore, we from now on, study the logarithmic function $f(x,y)$ instead of the density function $\rho(x,y)$:

$$f(x,y) = \ln \rho(x,y). \qquad (2.4)$$

2.2 Description of a Net of Orthogonal Curves

The following equation

$$\begin{cases} x = x(\lambda, \mu) \\ y = y(\lambda, \mu) \end{cases} \qquad (2.5)$$

tells us how to describe a net of curves by employing mathematics. Given two functions, $x(\lambda, \mu)$, $y(\lambda, \mu)$, you have the transformation between the curvilinear coordinates (λ, μ) and the rectangular Cartesian coordinates (x, y), i.e., the formula (2.5). It describes a net of curves. Letting the second parameter μ be a constant, you have a curve (called a row curve, or a proportion row). That is, the formula (2.5) is a curve with its parameter being λ. For the different values of the constant μ, you have a set of "parallel" rows. On the other hand, you let the first parameter λ be a constant then you have another curve (called a column curve, or a proportion column). That is, the formula (2.5) is a curve with its parameter being μ. For the different values of the constant λ, you have a set of "parallel" columns.

However, The row curves and the column curves are not necessarily orthogonal to each other. The following equation is the necessary and sufficient condition for the net of curves to be orthogonal:

$$\frac{\partial x}{\partial \lambda}\frac{\partial x}{\partial \mu} + \frac{\partial y}{\partial \lambda}\frac{\partial y}{\partial \mu} \equiv 0. \tag{2.6}$$

To study the rational structures described in the following, we need more knowledge of the description of row and column curves. The arc length of the row curve is s whose differential is the following:

$$ds = \sqrt{x_\lambda'^2 + y_\lambda'^2}d\lambda = Pd\lambda. \tag{2.7}$$

where P is the arc derivative of the row curve:

$$P = s_\lambda' = \sqrt{x_\lambda'^2 + y_\lambda'^2}. \tag{2.8}$$

The arc length of the column curve is t whose differential is the following:

$$dt = \sqrt{x_\mu'^2 + y_\mu'^2}d\mu = Qd\mu. \tag{2.9}$$

where Q is the arc derivative of the column curve:

$$Q = t_\mu' = \sqrt{x_\mu'^2 + y_\mu'^2}. \tag{2.10}$$

2.3 The Condition of Rational Structure

The formulas (2.5) and (2.6) are the general description of a net of orthogonal curves, and the formula (2.1) is the general description of a distribution of densities (a structure). This book talks about rational structure. A distribution of densities is called the rational structure if its density varies proportionally along some particular net of orthogonal curves. That is, you walk along a curve from the net and the ratio of the density on your left side to the immediate density on your right side is constant along the curve. However, the constant ratio from this curve is generally different from the constant ratios from other curves.

We have shown that a logarithmic ratio of two densities divided by the distance between the two positions is approximately the directional derivative of the logarithmic density along the direction of the distance. Therefore, we always study the logarithmic function $f(x, y)$ (see the formula (2.4)). If we know the two partial derivatives $\partial f / \partial x$ and $\partial f / \partial y$ then the structure $f(x, y)$ is found. The partial derivatives of $f(x, y)$ are the directional derivatives along the straight directions of the rectangular Cartesian coordinate lines. However, we are interested in the net of orthogonal curves and what we look for is the directional derivatives along the tangent directions of the curvilinear rows and columns. These are denoted $u(\lambda, \mu)$ and $v(\lambda, \mu)$ respectively.

The condition of rational structure is that u depends only on λ while v depends only on μ:

$$
\begin{aligned}
u &= u(\lambda), \\
v &= v(\mu).
\end{aligned}
\qquad (2.11)
$$

What a simple condition for the solution of rational structure!

Now we prove the condition. Assume you walk along a row curve. The logarithmic ratio of the density on your left side to the immediate density on your right side is approximately the directional derivative of $f(x, y)$ along the column direction. That is, the logarithmic ratio is approximately the directional derivative $v(\lambda, \mu)$. Because $v(\lambda, \mu)$ is constant along the row curve (rational), $v(\lambda, \mu)$ is independent of λ: $v = v(\mu)$. Similarly, we can prove that $u(\lambda, \mu) = u(\lambda)$.

2.4 Rational Structure Equation

It is known that, given an arbitrary function, we can have its two partial derivatives. However, given two functions, we may not find the third function whose partial derivatives are the given functions. For the given two functions to be some derivatives, a condition must be satisfied. The condition is the Green's theorem. In the case of derivatives along orthogonal curves, the Green's theorem is the following:

$$\frac{\partial}{\partial \mu}(u(\lambda, \mu)P) - \frac{\partial}{\partial \lambda}(v(\lambda, \mu)Q) = 0 \qquad (2.12)$$

In the case of rational structure, directional derivatives are the functions of the single variables, λ and μ respectively (see (2.11)). Therefore, the Green's theorem turns out to be the following which is called the rational structure equation:

$$u(\lambda)P'_\mu - v(\mu)Q'_\lambda = 0 \qquad (2.13)$$

This equation determines rational structure. To find a rational structure, generally we are first of all given a net of orthogonal curves. Accordingly the arc derivatives of both the rows and columns, $P(\lambda, \mu)$ and $Q(\lambda, \mu)$, are known. Therefore, the remaining functions , $u(\lambda)$ and $v(\mu)$, are the only unknowns. Note that the unknowns are the functions of the single variables, λ and μ respectively. Because the rational structure equation involves no derivative of the unknowns, the equation is not a differential equation at all. It is an algebraic equation and what we need to do is to add two factors (the two unknowns) to the derivatives of P and Q so that the rational structure equation holds: factorization!

How simple the universe is constructed!

2.5 The First Example: Spiral Galaxy Disk Structure

Images of spiral galaxies taken with infrared light show that each spiral galaxy is mainly a disk (the circularly symmetric disk with respect to the galaxy center, i.e., the disk center) and the disk light density decreases exponentially outwards along the radial direction from the center. There are other minor or weak structures in spiral galaxies. Spiral galaxies gain their name by

15

the fact that they present more or less spiral structures, known as arms.

The first example of rational structure is the spiral galaxy disk and is determined by the following net of equiangular spirals (or called logarithmic spirals, see Fig. 2.1):

$$\begin{cases} x = e^{d_1\lambda + d_2\mu} \cos(d_3\lambda + d_4\mu), \\ y = e^{d_1\lambda + d_2\mu} \sin(d_3\lambda + d_4\mu) \end{cases} \tag{2.14}$$

where $d_1(> 0), d_2(> 0), d_3(< 0), d_4(> 0)$ are real constants and we choose

$$d_3 = -d_1 d_2/d_4 \tag{2.15}$$

so that the curves are orthogonal. The polar angle and polar distance of the point (x, y) are easily seen,

$$\begin{aligned} r &= e^{d_1\lambda + d_2\mu}, \\ \theta &= d_3\lambda + d_4\mu. \end{aligned} \tag{2.16}$$

The coordinate lines are spiral-shaped and shown in the figure.

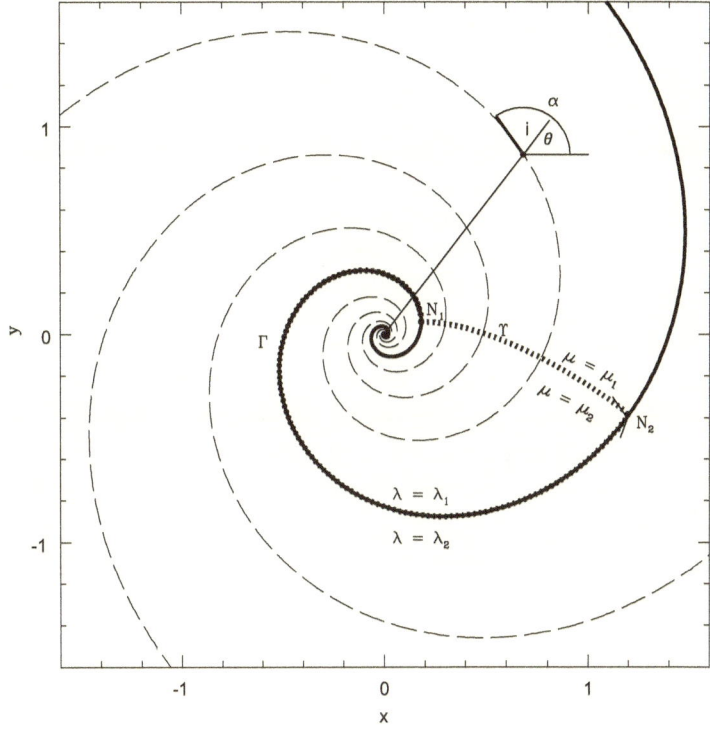

Figure 2.1: The orthogonal net of logarithmic spirals (2.14). The angle i at each position on the spiral between its bending direction and the disk radial direction, is constant along the curve. The closed curve which consists of two sections of equiangular spirals (thick dotted lines), demonstrates the closure condition (see the Appendix A for details).

The corresponding arc-length derivatives P, Q are

$$P(\lambda, \mu) = s'_\lambda = (d_1/d_4)\sqrt{d_2^2 + d_4^2}e^{d_1\lambda+d_2\mu},$$
$$Q(\lambda, \mu) = t'_\mu = \sqrt{d_2^2 + d_4^2}e^{d_1\lambda+d_2\mu}. \qquad (2.17)$$

The rational structure equation (2.13) does help factor out the required directional derivatives for our spiral galaxy disks,

$$u(\lambda) = d_5 d_4,$$
$$v(\mu) = d_5 d_2 \qquad (2.18)$$

where d_5 is another constant.

So far, we did not specify the variance domain S on (λ, μ) coordinate plane on which the coordinate system (2.14) is defined. There are many such domains which are proved in the Appendix of the book. Here we simply present one kind of such domain. We define a constant,

$$\Delta_\lambda = \frac{2\pi d_2}{d_1 d_4 - d_3 d_2} \ (> 0), \qquad (2.19)$$

The domain is the following:

$$S_{\lambda_1} : \lambda_1 < \lambda < \lambda_2(= \lambda_1 + \Delta_\lambda), \ -\infty < \mu < +\infty. \qquad (2.20)$$

where λ_1 is arbitrary constant and the length of the interval (λ_1, λ_2) is

$$\Delta_\lambda = \lambda_2 - \lambda_1. \qquad (2.21)$$

In the Appendix, we also prove that the spiral curves are equiangular spirals.

The spiral disk density along all orthogonal coordinate lines can be found by performing the path integrations of the following formulas

$$df = uds = u(\lambda)Pd\lambda,$$
$$df = vdt = v(\mu)Qd\mu. \qquad (2.22)$$

along the row curves, $\mu = $ constant, and the column curves, $\lambda = $ constant, respectively. Without loss of generality, we assume the coordinates λ and μ are defined on the domain (2.20). Then

18

starting at the galaxy center, we perform the path integration of $df = v(\mu)Qd\mu$ over $[-\infty, \mu]$ by taking λ to be the constant λ_1 to get $f(\lambda_1, \mu)$. Then we perform the path integration of $df = u(\lambda)Pd\lambda$ over $[\lambda_1, \lambda]$ by taking μ to be an arbitrary constant to get $f(\lambda, \mu)$. Finally, we have the logarithmic function $f(x, y)$ implied by the spiral-shaped coordinate system. The density distribution $\rho(x, y)$ represents spiral galaxy disks (we choose $d_5 < 0$, because light density $\rho \to 0$ when $r \to +\infty$),

$$
\begin{aligned}
f_d &= d_5\sqrt{d_2^2 + d_4^2}\,e^{d_1\lambda + d_2\mu}, \\
\rho_d &= d_0 \exp(d_5\sqrt{d_2^2 + d_4^2}\,e^{d_1\lambda + d_2\mu})
\end{aligned}
\tag{2.23}
$$

where d_0 is the light density (star density) at the galaxy center. Note that we use the letter d as well as the subscript d for disk parameters and formulas. Similar notations are used for bar parameters and formulas. Because $f_d = d_5 Q(\lambda, \mu)$ and $Q(\lambda, \mu)$ is uniquely defined over the whole galaxy disk plane ((x, y) plane, see the Appendix A), $f_d(x, y)$ is uniquely defined on the same plane. We can see the disk light pattern by displaying $\rho_d(x, y)$. Because the polar distance is $r = \exp(d_1\lambda + d_2\mu)$, the galaxy disk light distribution is circularly symmetric,

$$
\rho_d = d_0 e^{f_d} = d_0 e^{(d_5\sqrt{d_2^2 + d_4^2})r}.
\tag{2.24}
$$

Since galaxy light density $\propto \rho$, we have recovered the known exponential law of spiral galaxy disk (the exponential disk).

The circularly symmetric exponential disk is completely determined by the value of d_0 and the value of $d_5\sqrt{d_2^2 + d_4^2}$. Therefore, given a spiral galaxy disk, that is, given the two specific values, we can find an infinite number of coordinate systems (2.14) defined on the disk plane which give the same disk structure. From now on, we choose $\sqrt{d_2^2 + d_4^2} = 1$ so that the disk model involves only two variable disk parameters $d_0(> 0)$ and $d_5(< 0)$,

$$
\rho_d = d_0 e^{f_d} = d_0 e^{d_5 r}.
\tag{2.25}
$$

In fact, galaxy structures depend only on the geometric curves, not the choice of coordinate parameters. Therefore, the following is the general expression for equiangular spirals:

$$\begin{cases} x = e^{d_1 f(\lambda) + d_2 g(\mu)} \cos(d_3 f(\lambda) + d_4 g(\mu)), \\ y = e^{d_1 f(\lambda) + d_2 g(\mu)} \sin(d_3 f(\lambda) + d_4 g(\mu)) \end{cases} \quad (2.26)$$

where $f(\lambda), g(\mu)$ are arbitrary functions. All these expressions give the same equiangular spirals and generate the same exponential disks, independent of the choice of coordinate parameters. That is, our rational structure which depends only on the geometric proportion curves, is independent of the coordinate system which describes the curves (coordinate invariance).

2.6 The Second Example: Dual Handle Structure

Now we study galactic bar model. A bar pattern is composed of two or three dual handle structures which have different lengths and are generally aligned with each other (spiral galaxy NGC 1365 is not the case, see Fig. 3.1). The dual handle structure is determined by the following orthogonal curves of confocal ellipses and hyperbolas:

$$\begin{cases} x = e^{\sigma} \cos \tau, \\ y = \sqrt{e^{2\sigma} + b_1^2} \sin \tau, \\ -\infty < \sigma < +\infty, \ 0 \le \tau < 2\pi \end{cases} \quad (2.27)$$

where $b_1 (> 0)$ is a constant. The orthogonal coordinate system no longer shares the coordinate lines with the polar coordinate system. The coordinate lines are confocal ellipses and hyperbolas (Fig. 2.2). The distance between the two foci is $2b_1$ which measures the distance between the two handles. The eccentric anomaly of the ellipses is τ.

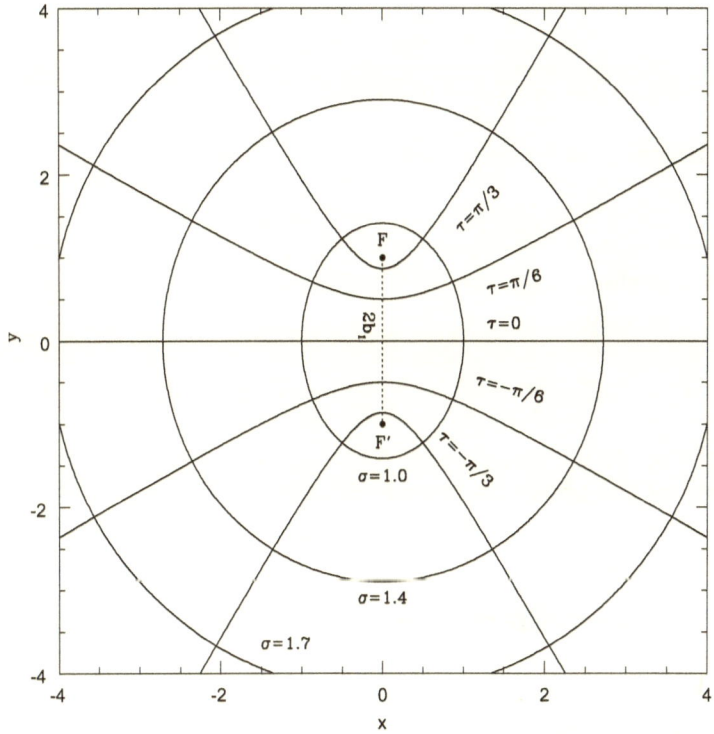

Figure 2.2: The orthogonal curves of confocal ellipses and hyperbolas. The distance between the two foci, F and F', is $2b_1$ which measures the distance between the two handles.

Now we look for the logarithmic density $f(\sigma, \tau)$ of the dual handle structure determined by the coordinate system (2.27). The corresponding arc-length derivatives of the coordinate system are (see (2.8) and (2.10)),

$$
\begin{aligned}
P &= e^\sigma \sqrt{e^{2\sigma} + b_1^2 \cos^2 \tau} / \sqrt{e^{2\sigma} + b_1^2}, \\
Q &= \sqrt{e^{2\sigma} + b_1^2 \cos^2 \tau}.
\end{aligned}
\tag{2.28}
$$

The rational structure equation

$$
u_b(\sigma) P'_\tau - v_b(\tau) Q'_\sigma = 0
\tag{2.29}
$$

determines the corresponding directional derivatives of the logarithmic density,

$$
u_b(\sigma) = b_2 e^\sigma \sqrt{e^{2\sigma} + b_1^2}, \quad v_b(\tau) = -b_2 b_1^2 \sin \tau \cos \tau.
\tag{2.30}
$$

To get the logarithmic density, we need to perform path integrations of the similar formulas to (2.22). The result is

$$
f_b(\sigma, \tau) = (b_2/3)(e^{2\sigma} + b_1^2 \cos^2 \tau)^{3/2}.
\tag{2.31}
$$

The inverse coordinate transformation of the formulas (2.27) is easily found,

$$
\begin{aligned}
p(x, y) &= e^\sigma = \sqrt{(r^2 - b_1^2 + \sqrt{(r^2 - b_1^2)^2 + 4b_1^2 x^2})/2}, \\
\cos \tau &= xe^{-\sigma} = x/p(x, y)
\end{aligned}
\tag{2.32}
$$

where $r^2 = x^2 + y^2$. Finally we find the density of the dual handle structure,

$$
\begin{aligned}
f_b(x, y) &= (b_2/3)(p^2(x, y) + b_1^2 x^2/p^2(x, y))^{3/2}, \\
\rho_b &= b_0 \exp(f_b(x, y))
\end{aligned}
\tag{2.33}
$$

where b_0 is the dual handle density at the galaxy center. We need to choose $b_2 < 0$ so that $f_b < 0$ and $\rho_b \to 0$ when $r \to +\infty$. We can see that b_0 corresponds to the central dual handle strength and b_1 corresponds to the dual handle length while b_2

measures the density slope off the dual handle. If we display the dual handle structure as a curved surface in 3-dimensional space then we can see that the surface is camelback-like shapes with two humps (i.e., handles).

Note that polar angle θ is not defined at the center $r = 0$ for the polar coordinates. Similarly the eccentric anomaly τ is not defined for the coordinates (2.27) along the dual handle central line of $2b_1$ length which is the coordinate line $\sigma = -\infty$.

In fact, galaxy structures depend only on the geometric curves, not the choice of coordinate parameters. Therefore, the following is the general expression for dual handle structures:

$$\begin{cases} x = e^{f(\sigma)} \cos g(\tau), \\ y = \sqrt{e^{2f(\sigma)} + b_1^2} \sin g(\tau) \end{cases} \tag{2.34}$$

where $f(\sigma), g(\tau)$ are arbitrary functions. All these expressions give the same dual handle structures, independent of the choice of coordinate parameters. That is, our rational structure which depends only on the geometric proportion curves, is independent of the coordinate system which describes the curves (coordinate invariance).

Chapter 3

The Origin of Spiral Galaxies

3.1 Prop. 1: Rational Structures are at Most Bilaterally Symmetric

This is a mathematical proposition: any net of orthogonal curves is either circularly symmetric with respect to the center point, or bilaterally symmetric. I spent three years from 2002 to 2005 to look for a net of orthogonal curves whose shape has odd symmetry. That is, I wanted to find a rational structure which resembled a two-arm spiral pattern like the spiral galaxy M51. The three-year study indicates that a net of orthogonal curves is generally connected to some complex analytical function. As you might know, a complex analytical function always has two parts (real and imaginary) like the formula (2.5), and leads to a net of orthogonal curves. However, Complex analytical functions are very special ones which satisfy some strong conditions like Cauchy integral theorem and formula. From my experience, I do not find any complex analytical function whose graph of the real or imaginary part has odd symmetry. Therefore, I have the proposition that rational structures are at most bilaterally

25

symmetric. Please help me prove the proposition.

Surprisingly, galaxy structures happen in the same way as indicated in the following.

3.2 Coinc. 1: Galaxy Patterns (except Arms) are at Most Bilaterally Symmetric

Amazingly, any component of any galaxy pattern (except the arm pattern) is either circularly symmetric, or bilaterally symmetric. And my academic papers [4-11] show that, except the arm structure, any component of any galaxy structure (such as exponential disks, galactic bars, even the whole elliptical galaxy) is a rational structure. That is, any galaxy image (ignoring the arm) can be fitted identically to rational structures.

The following will present more and more cases of coincidences. Therefore, they are not coincidence at all. They are the cosmic truth, and Dr. He's discovery is absolutely important!

3.3 Coinc. 2: Dust and Irrationality

It is the observational fact that spiral galaxies are full of dust while elliptical galaxies have no dust. Why does this happen? However, people have not found its answer since galaxies were discovered more than 80 years ago. Dr. He gives the answer.

Arm structure is neither circularly symmetric with respect to the center point, nor bilaterally symmetric. Therefore, arms are not rational structures. Arm pattern tends to be oddly symmetric with respect to the center. But we can never find such "grand design" arm pattern which presents the perfectly odd symmetry. In fact, there are very different types of spiral structures. Some galaxies (like M51, Fig. 1.1) are what we call "Grand Design" spirals, meaning that they have a clearly outlined and well organized spiral structure. Other galaxies, like NGC 4414 are called "flocculent" and have much harder to trace arms. Compared with the exponential disks, galactic bars, and even the whole elliptical galaxies which demonstrate smooth and rational structures, arms may not be called a structure at all.

Arm patterns exist only in spiral galaxies and they are weak compared with the main disk structure of spiral galaxies. Therefore, the presence of arm structure is the disturbance to the rational structure. Because arm patterns exist only in spiral galaxies and only spiral galaxies present dust, Dr. He comes to the critical answer to the above question: any disturbance to rational structure must produce cosmic dust!

3.4 The Origin of Cosmic Dust: Impulsive Gravity

From now on we call the disturbance to rational structure the cosmic impulse. Therefore, the universe is originated not only rationally but also impulsively! In the case of large-scale structures (that is, galaxies), the impulse is demonstrated to be the disturbing waves, i.e., the arm patterns.

We have seen that the rational structure is the main structure while the disturbance to the main structure is always weak. That is, rational force is the main one while the impulse is weak. In the case of large-scale structures (that is, galaxies), the disturbing waves try to achieve the minimal disturbance and, as a result, they follow the proportion rows or columns of the rational structures. The impulsive disturbance reveals the rational design of the universe: proportion!

Because we live inside a galaxy (Milky Way galaxy) and we find no other force exists except gravity at our neighborhood, we conclude that Newton or Einstein formulation of gravity is a partial result of the universal "gravity" and the real gravity presents not only rational sense but also impulsive one as indicated at the large-scale galaxy structures!

3.5 A Simpler Argument that Spiral Galaxy Disk is a Rational Structure

The logarithmic stellar density of exponential disk is the circularly symmetric distribution of numerical values about the center point, which decreases linearly in the radial directions. Therefore, the magnitude of the directional derivatives along radial directions is a global constant. An equiangular spiral is the curve which makes a constant angle to the local radial directions. Therefore, the magnitude of the directional derivatives along the tangent directions of an equiangular spiral is constant along the spiral. Accordingly the magnitude of the directional derivatives along the perpendicular directions to the equiangular spiral is also constant along the spiral. That is, you walk along an equiangular spiral and the ratio of the density on your left side to the immediate density on your right side is constant along the curve. However, this constant is different from the constant in the radial direction. They differ by a factor of the cosine of the angle. Finally we have proved that exponential disk is a rational structure.

All curves which are orthogonal to a specific set of equiangular spirals in the clockwise direction are themselves equiangular but in the counter-clockwise direction. These two sets of spirals in opposite directions consist of the orthogonal net of curves, and the exponential disk is the rational structure with respect to these proportion curves.

3.6 Coinc. 3: Exponential Disk is Correlated with Equiangular Spiral

The exponential disk of any spiral galaxy is a rational structure which is circularly symmetric about the galaxy center. It

28

is amazing that its proportion curves are the equiangular spirals which are precisely the curves represented by normal spiral galaxy arms. This is another coincidence.

Now we introduce more and more coincidences.

3.7 Coinc. 4: The Only Rational Structure which is not Circularly Symmetric is Dual Handle Structure

I have given the definition of rational structure. However, given an arbitrary net of orthogonal curves, we are not always possible to arrange a distribution of stellar density on the net to form a rational structure. In fact, there are only a few types of orthogonal curves which correspond to rational structures. Rational structures are usually circularly symmetric about the center points. The only rational structure we can find which is not circularly symmetric, is the bilaterally symmetric structure, namely, dual-handle structure.

We have two types of rational structures: exponential disks and dual-handle structures. Adding the two structures together leads to the barred pattern as we expected! It is amazing that only two kinds of spiral galaxies are observed in the universe. One kind of spirals are the normal spiral galaxies while the other kind are the barred spiral ones. What is more surprising is that some barred galaxies do show a set of symmetric enhancements at the ends of the stellar bar, called the ansae or the "handles" of the bar (see upper-left panel of Fig. 3.1). This indicates that a bar itself is nothing but a set of several pairs of ansae (handles). That is, bars are the superposition of several aligned or misaligned dual-handle structures. If the outer dual-handle structure is far more away from the galaxy center then it demonstrates the pattern of ansae or "handles" of the bar.

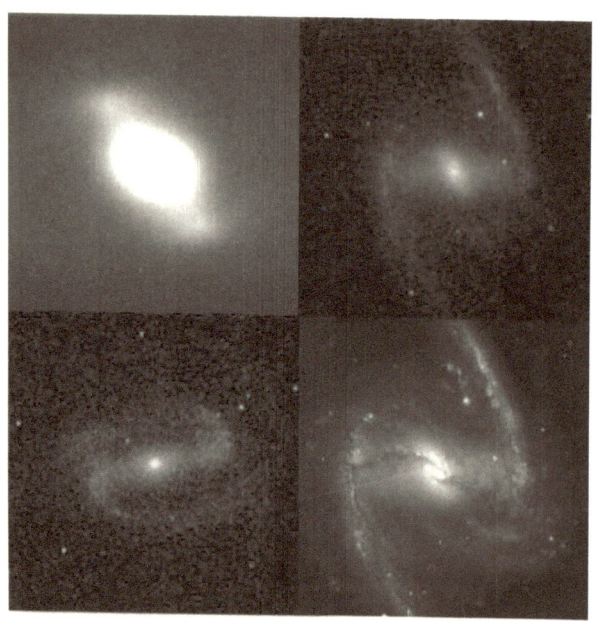

Figure 3.1: Upper-left panel is galaxy NGC 2983 (see Co-inc. 4 for its explanation, image credit [12]). Upper-right is the infrared image of galaxy NGC 1365 (see Coinc. 5, image credit [1]). Lower-left is the infrared image of NGC 1300 (see Coinc. 7, image credit [1]). Lower-right is NGC 1365 (ultraviolet image, credit: European Southern Observatory).

3.8 Coinc. 5: There are Barred Spiral Galaxies which Present Two Nonparallel Bars

The main structure of spiral galaxies is the exponential disk. When the dual-handle structure (i.e., sub-bar) is near the galaxy center, the superposition of the dual-handles to the bright disk center presents a bar shape. This precisely explains the origin of galaxy bars. A galaxy bar is usually composed of two or more sets of aligned or misaligned dual-handle structures. Surprisingly, there are barred spiral galaxies which present two nonparallel bars (see upper-right panel of Fig. 3.1).

3.9 Coinc. 6: Bar Structure is So Weak in the Outer Areas of Spiral Galaxies that it is Ignored

Compared with the exponential disk, the bar is observationally weak structure. That is, bar structure is so weak in the outer areas of spiral galaxies that it is ignored. It is very surprising that the theoretical calculation of dual-handle structure shows that it is weak when compared with the disk (see Fig. 3.2, 3.3, and 3.4)!

We know that the disk density of spiral galaxies decreases outwards exponentially, which is the numerical result obtained over 90 years since the discovery of galaxies in the universe. Spiral galaxy disks are thus called exponential disks. We add the dual-handle structure to the exponential disk for them to be the model of barred spiral galaxies. If the density of dual-handle structure were comparable to or stronger than the exponential disk in the far distances from the galaxy center then our model would fail. That would suggest that the main structure of spiral galaxies were not the exponential disk, a result inconsistent with astronomical observation. The mathematical result is that the density distribution of dual-handle structure decreases outwards cubic-exponentially. In the distances far from the galaxy center, the dual-handle structure is negligible when compared with the disk. Mathematical result is consistent to observation! This piece of coincidence alone proves that galaxies are originated from proportion force!

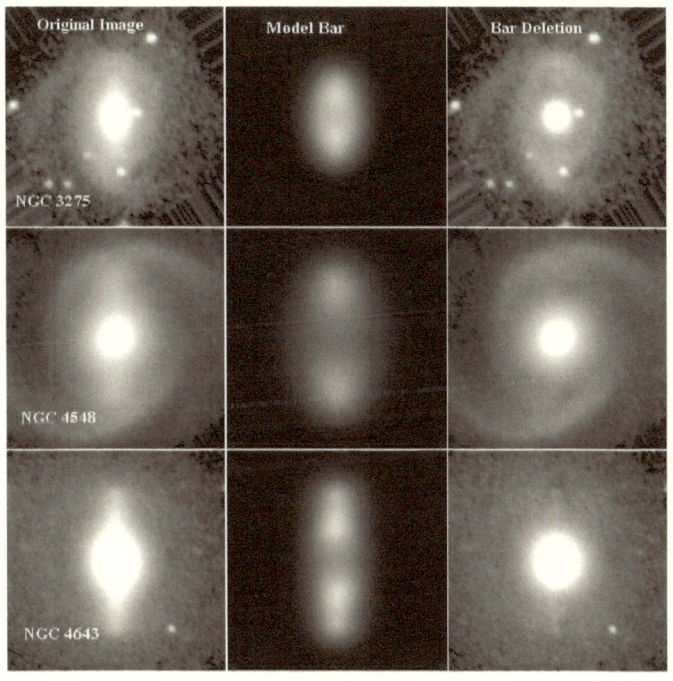

Figure 3.2: The OSUBGS H-band images NGC 3275, 4548, 4643 (image credit [13]) minus our model bars respectively result in the disk and bulge images (bar deletion).

3.10 Coinc. 7: The Arms of Barred Galaxies Spin around the Bar and are No Longer the Equiangular Spirals

With simple mathematical calculation we know that spiral-shaped proportion curves exist in dual-handle structure. However, they are not equiangular because they surround the central line of the dual-handles (recall that the spirals in exponential disks are equiangular and surround the center point). Two proportion curves which are oddly symmetric about the center point in dual-handle structure make approximately elliptical shape and its long axis must be parallel to the central line of the dual-handles. Surprisingly, astronomical observations show that arms of barred spiral galaxies do surround the middle lines of their bars, and they are not equiangular, and the two arms make approximately elliptical shapes with the long axes being parallel to the bar middle lines (see Fig. 3.1).

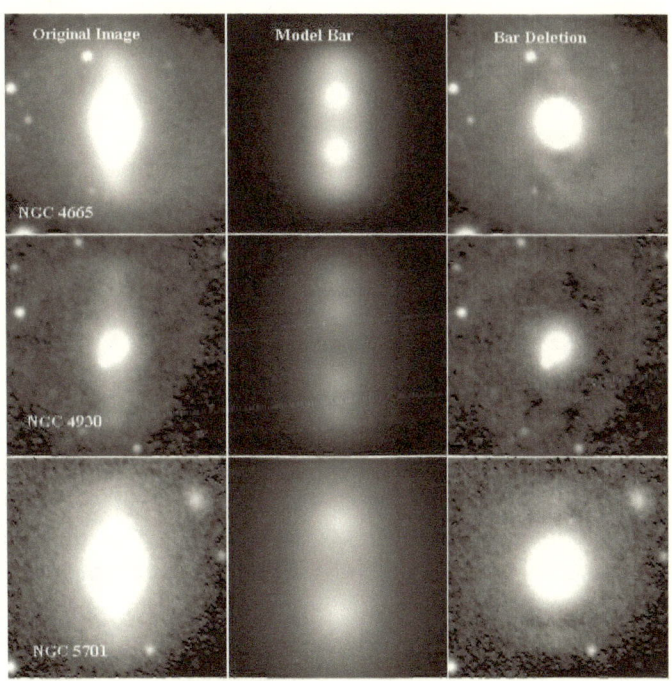

Figure 3.3: The OSUBGS H-band images NGC 4665, 4930, 5701 (image credit [13]) minus our model bars respectively result in the disk and bulge images (bar deletion).

We proceed with more mathematical details. The proportion curves of a dual-handle structure are all confocal ellipses and hyperbolas. The two foci are the centers of the two handles. The distance between the two foci is the length of the dual-handle structure.

If the length of the dual-handle structure is zero then the two foci overlap to be the center of concentric circles and the above-said proportion curves become the curves of polar coordinates. This returns to the case of normal spiral galaxy disk. In fact, the polar curves are also the proportion curves of normal spiral galaxies. In other words, all polar curves are the limiting curves of equiangular spirals.

Back to the above-mentioned dual-handle structures. Similarly, they have also open proportion spirals which make acute angles to the above-mentioned ellipses and hyperbolas, and spin around the central line of the dual-handle structure. However, the spirals are no longer equiangular.

3.11 Coinc. 8: Circular and Elliptical Rings

We have already known that exponential disks have circular proportion curves (one family of polar curves). Observationally, some normal spiral galaxies do have closed arms which are circular, called rings. Dual-handle structures also have closed proportion curves which are ellipses whose long axes must be parallel to the central lines of the dual-handles. Observationally, some barred spiral galaxies do have closed rings which are ellipses and the long axes are parallel to the galaxy bars.

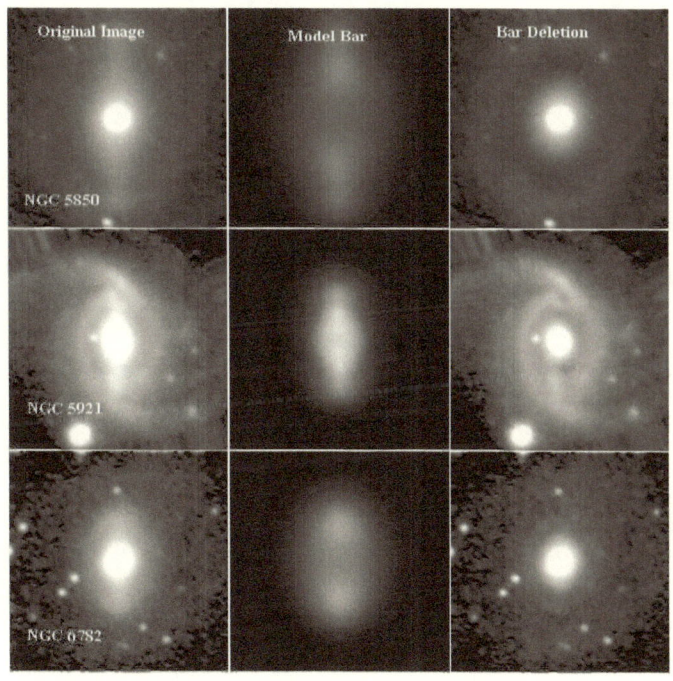

Figure 3.4: The OSUBGS H-band images NGC 5850, 5921, 6782 (image credit [13]) minus our model bars respectively result in the disk and bulge images (bar deletion).

3.12 Coinc. 9: Fitting Bar Images with Dual-handle Structures

In fact, we can directly use dual-handle structures to fit into the bar images of spiral galaxies. From the barred spiral galaxy images we subtract the superposition of two or three sets of dual-handle structures and see if the resulting images are exponential disks plus the weak arm structures. If it works then we have further proved that barred spiral galaxies are the superposition of exponential disks with dual-handle structures. The first column of the Figures 3.2, 3.3 and 3.4 present real galaxy images and the second column present the fitting bars. The third column presents the result of the subtraction of the second column from the first one. The results are very good: the resulting images are exponential disks plus the weak arm structures. In fact, Dr. He has made a piece of computer software which can be used to improve the results.

3.13 Coinc. 10: Elliptical Galaxies are Completely Rational Structures

The next Chapter proves that elliptical galaxies are completely rational structures in three-dimensions. The proportion surfaces of elliptical galaxies are the intersecting nets of orthogonal spheres, where disturbing waves are difficult to form and spread. On the other hand, spiral galaxies are two-dimensional and their proportion curves are open spirals where disturbance waves are easy to form and spread. Astronomical observations do show that arms do not exist in elliptical galaxies.

The disturbance to rational structure leads to the formation of gas and dust. New families of stars and planets are born to these gas and dust. The star-planet families are short-lived. This happens only in spiral galaxies.

3.14 It Seems that Fitting Galaxy Images may Tell the Physical Distances of the Corresponding Galaxies

Galaxies are very far away from the earth and human is really minute. Therefore, we can not measure directly how far away the galaxies are from earth. It seems that fitting barred galaxy images with dual-handle structures may tell the physical distances of the corresponding galaxies. However, the result needs further confirmation with the help of supercomputers.

Chapter 4

The Origin of Elliptical Galaxies

4.1 Introduction

Compared to disk galaxies, elliptical galaxies are more "clean" objects for people to test models on the galactic-scale gravitational systems because they have smoother texture of the photographic images. Their light distributions can be approximately the linear representation of their mass distributions. Real elliptical galaxies are 3-dimensional distributions of light. The observed 2-dimensional elliptical-shaped light distributions (i.e., surface brightness) result from the integration of the light along the line-of-sight, i. e., the projection on to the 2-dimensional sky plane. Therefore, a complete understanding of elliptical galaxy patterns is the analytical formulation of their 3-dimensional light distributions, i. e., the analytical deprojection of the observed 2-dimensional profiles. The 3-dimensional light distributions have two aspects: one is the law of their one dimensional radial density profile and the other is their 3-dimensional shapes.

Fortunately, the projected one dimensional radial profile is studied very clear. It decreases smoothly from the nucleus, closely following the Sérsic law [14]

$$\Sigma_s(R) = \sigma_s \exp(\sigma_0 R^{1/n}) \tag{4.1}$$

where $\sigma_s(> 0)$ is the constant surface brightness at the galaxy center, $\sigma_0(< 0)$ is another constant, and R is the projected radial distance from the galaxy center along the semi-major axis. However, the spatial (3-dimensional) radial distance from the galaxy center is denoted, from now on, by r. The Sérsic law, which fits galaxy inner parts very well, indicates that galaxy light gradients at their centers are $-\infty$. The Hubble law [15] suggests an unrealistic zero-gradient at the galaxy centers (flattened light distribution). Much attention has been given to the deprojection of Sérsic law, that is, the proposition of the spatial (3-dimensional) light distributions (or equivalently, in the case of spherically symmetric light distributions, the proposition of the spatial one dimensional radial profile) whose projected line profiles follow the Sérsic law. However, the deprojection of the Sérsic law is not tractable analytically ([16]). An analytical formula of the 3-dimensional light distributions whose projected line profile resembles the $R^{1/n}$ Sérsic law or the $R^{1/4}$ de Vaucouleurs law (i.e., Sérsic law when $n = 4$) is of great interest. The Dehnen model ([17]) is about spherically symmetric light distributions, and resembles the $R^{1/n}$ Sérsic law when $n = 4$,

$$\rho(r) = \frac{b}{r^\gamma} \frac{1}{(r + a)^{4-\gamma}} \tag{4.2}$$

where $a(> 0), b(> 0)$ are constants.

Another aspect of elliptical galaxy study is their shapes. Elliptical galaxies gain their name by their projected apparent shapes. However, their 3-dimensional shapes, rotationally symmetric or triaxial, are hard to infer. The assumption of triaxial shapes is extensively used. Stark ([18]) proposed a model of triaxial ellipsoid isophotes of constant axis ratios whose projected isophote curves are exact ellipticals of constant axis ratios too. The isophote curves of real images, however, have radially varying axis ratios. Carter ([19])'s investigation indicates that ellipticity of the isophotes is an increasing or peaked function of radius. In addition, the isophote curves are not exact elliptical. They are sometimes boxy-shaped and sometimes disky-shaped ([20]).

4.2 My Rational Model of Elliptical Galaxies

There must be some simple symmetry behind the Sérsic or Dehnen law of elliptical galaxies and their shapes of three dimensions. My proposition is the rational gravity. The proportion symmetry maps a net of orthogonal curves to a logarithmic density distribution. For spiral galaxy disk, the net of orthogonal curves is the formula (2.14), which is formed by the complex analytical function $Z = e^W$. A "disturbed" form of the orthogonal curves is the formula (2.27) which is used to model galactic bars. In the present Chapter, I generalize the idea to 3-dimensional density distributions and present a model for elliptical galaxy patterns. An orthogonal coordinate system is employed which is formed by the complex plane transformation $Z = 1/W$. The resulting spatial (3-dimensional) light distribution is spherically symmetric and has infinite gradient at the center, which is called spherical-nucleus solution and is used to model galaxy central areas. We can make changes of the coordinate system by cutting out some column areas of its definition domain, the areas containing the galaxy center. The resulting spatial (3-dimensional) light distributions are rotationally symmetric or triaxial and have zero gradient at the center, which are called elliptical-shape solutions and are used to model galaxy global elliptical structures. The two kinds of logarithmic light distributions are added together to model the full elliptical galaxy patterns. The model is the generalization of Dehnen's model ([17]) whose projected profile resembles de Vaucouleurs' law. One of the elliptical-shape solutions permits realistic numerical calculation and is fitted to all R-band elliptical images from Frei *et al.* ([21])'s galaxy sample. The model fits real galaxy shapes satisfactorily. This suggests that elliptical galaxy patterns can be represented in terms of a few basic parameters.

4.3 A Model of Galaxy Central Areas: the Spherical-nucleus Solution

(i) *Spherically Symmetric Density Distribution.* Spherically symmetric structures are also rotationally symmetric:

$$f(r) \equiv f(x, y, z) \equiv f(R, z), \tag{4.3}$$

where $R = \sqrt{x^2 + y^2}$ and the z-axis is the axis about which the 3-dimensional density distribution (light distribution) is rotationally symmetric. Therefore, we first introduce the standard description of orthogonal curves defined on a 2-dimensional plane (R, z) which is any plane containing the z-axis (see Fig. 4.1). A curvilinear coordinate system defined on the plane labels each point (R, z) with an ordered set of two real numbers (λ, μ) over a region S on (λ, μ) plane,

$$\begin{aligned} R &= R(\lambda, \mu), \\ z &= z(\lambda, \mu). \end{aligned} \tag{4.4}$$

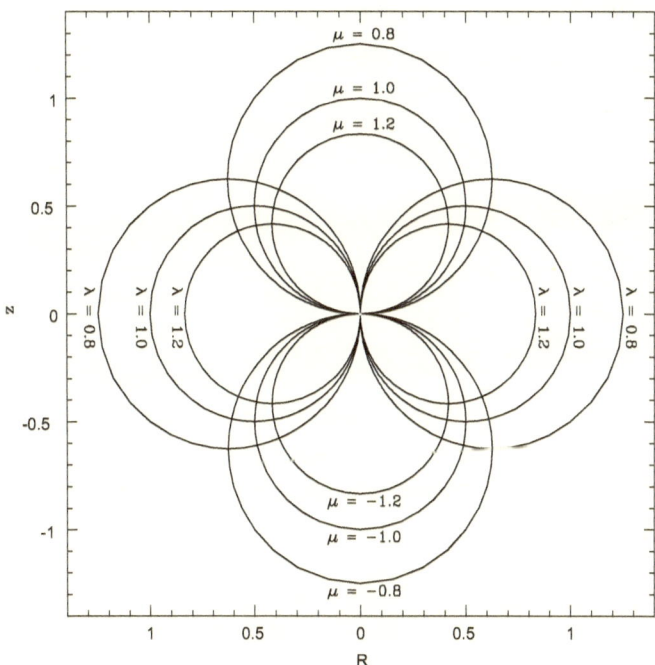

Figure 4.1: The orthogonal 4-circle coordinate system. The (R, z) plane is any one in 3-dimensional space which contains the Cartesian z-axis. Therefore, $R = \sqrt{(x^2 + y^2)}$.

(ii) *Complex Reciprocal Function.* Now we present the co-ordinate system on which all the models proposed in the Chapter are based,

$$R = \lambda/(\lambda^2 + \mu^2),$$
$$z = \mu/(\lambda^2 + \mu^2),$$
$$0 < \lambda < +\infty, \quad -\infty < \mu < +\infty. \tag{4.5}$$

The orthogonal coordinate system can be produced by the complex analytical function $Z = 1/W$ where $Z = R + iz$ and $W = \lambda - i\mu$. The corresponding arc-length derivatives P, Q are

$$P(\lambda, \mu) = s'_\lambda = 1/(\lambda^2 + \mu^2),$$
$$Q(\lambda, \mu) = t'_\mu = 1/(\lambda^2 + \mu^2) \equiv P. \tag{4.6}$$

The inverse formulas of the coordinate transformation are

$$\lambda = R/(R^2 + z^2),$$
$$\mu = z/(R^2 + z^2),$$
$$0 < R < +\infty, \quad -\infty < z < +\infty. \tag{4.7}$$

Now we look for the logarithmic density distribution $f(R, z)$ defined on the (R, z) plane. Its directional derivatives along the tangent directions of the curves can be found by the rational structure equation (2.13),

$$u(\lambda) = h\lambda, v(\mu) = h\mu \tag{4.8}$$

where h is constant. The logarithmic density along all orthogonal coordinate lines can be found by performing path integrations of the following formulas

$$df = uds = u(\lambda)Pd\lambda,$$
$$df = vdt = v(\mu)Qd\mu \tag{4.9}$$

along the coordinate lines $\mu = $ constant and $\lambda = $ constant respectively. Choosing the galaxy origin to be $R = 0, z = 0$ and performing the path integration, we can have our light distribution on the (R, z) plane, $f = (h/2)\ln(\lambda^2 + \mu^2)$. We

46

can prove the formula by verifying the relation between the physical components of its gradient, u, v, and the contravariant components, $\partial f/\partial\lambda, \partial f/\partial\mu$. The relation is, $u = (\partial f/\partial\lambda)/P$, $v = (\partial f/\partial\mu)/Q$. Using the inverse formulas of the coordinate transformation (4.7), we finally have our logarithmic and direct density distributions respectively,

$$
\begin{aligned}
&f(R, z) = (h/2)\ln(\lambda^2 + \mu^2) = (h/2)\ln \tfrac{1}{R^2+z^2}, \\
&\rho(R, z) = \rho_0(\tfrac{1}{R^2+z^2})^{h/2}, \\
&0 < R < +\infty, \; -\infty < z < +\infty.
\end{aligned}
\tag{4.10}
$$

where ρ_0 is another constant. We choose $h > 0$ because density $\rho = \rho_0 \exp f \to 0$ when $R, z \to +\infty$. Note that I have purposely neglected the integration constant f_0 on the right-hand side of the first equation in (4.10). Recovering the integration constant is equivalent to adding a constant factor besides $(\lambda^2 + \mu^2)$ or above $(R^2 + z^2)$ in (4.10). The legitimacy and the apparent error of the neglect of the integration constant are explained in the following.

Significance of Physical Dimensions. As a basic requirement for any model equation, the physical dimensions of the left and right-hand side must be identical and the argument of, e. g., a logarithm or an exponential must be a dimensionless number. However, the arguments of the logarithms in the first equation of (4.10) lack a constant factor besides $(\lambda^2 + \mu^2)$ or above $(R^2 + z^2)$. Because R and z have the dimension of length, the corresponding constant factor must have the inverse dimension of length so that the argument of the logarithm is a dimensionless number. Neglect of the factor appears wrong. However, the constant factor corresponds to an additional term, i. e. the integration constant, to the logarithmic distribution $f(R, z)$, and the term corresponds to an additional factor to the direct density distribution $\rho(R, z)$. If ρ_0 is literally considered to be galaxy density distribution then we need to find galaxy distance from earth in addition to its digital image light distribution. The distance, however, is very hard to determine while the difference between real galaxy light density and its image light density is

47

always a proportional factor. Therefore, I choose to incorporate this proportional factor as well as the above-said additional
factor into ρ_0, and consider the incorporated factor ρ_0 to be a
fitting parameter. That is, any galaxy image without the deter-
mination of its real galaxy light density is directly fitted to my
model light distribution with the fitting parameter ρ_0 as well as
other parameters. Similarly, on the right-hand side of the first
equation of (4.10), the dimensionless logarithmic distribution
$f(R, z)$ should have the other factor f_1 to achieve its complete
degrees of freedom. However, I choose to incorporate the factor
into $(h/2)$.

(iii) *Galaxy Structures Depend Only on the Geometric
Curves.* In fact, galaxy structures depend only on the geometric curves, not the choice of coordinate parameters. Therefore,
the following is the general expression for the density distribution (4.5):

$$\begin{cases} R = f(\lambda)/(f(\lambda)^2 + g(\mu)^2), \\ z = g(\mu)/(f(\lambda)^2 + g(\mu)^2). \end{cases} \tag{4.11}$$

where $f(\lambda), g(\mu)$ are arbitrary functions. All these expressions
give the same density distribution, independent of the choice
of coordinate parameters. That is, our rational structure which
depends only on the geometric proportion curves, is independent
of the coordinate system which describes the curves (coordinate
invariance).

(iv) **A Model of Galaxy Central Areas: the Spherical-
nucleus Solution.** We rotate the above (R, z) plane about
the z-axis (i.e. the line $R = 0$) to achieve our 3-dimensional
galaxy density distribution which is rotationally symmetric. The
resulting density distribution is

$$\begin{aligned} f_n(x, y, z) &= (h/2) \ln \frac{1}{x^2+y^2+z^2}, \\ \rho_n(x, y, z) &= \rho_0 (\frac{1}{x^2+y^2+z^2})^{h/2} = \rho_0 \frac{1}{r^h} \\ &-\infty < x, y, z < +\infty. \end{aligned} \tag{4.12}$$

We see that it is not only rotationally symmetric but also spherically symmetric. The density at the galaxy center is $+\infty$ and the
gradient of density at the center is $-\infty$. The solution is called

spherical-nucleus solution and is used to model galaxy central areas. In fact, it corresponds to the first factor of the Dehnen model (4.2) with γ being replaced by h. We can project the 3-dimensional distribution on any plane to see its surface brightness. We choose the x-y plane to be the sky plane. The surface brightness is

$$
\begin{aligned}
\Sigma(x,y) &= \int_{-\infty}^{+\infty} \rho(x,y,z)dz \\
&= 2\rho_0 \int_0^{+\infty} (\frac{1}{x^2+y^2+z^2})^{h/2} dz \\
&= 2\rho_0 \int_0^{+\infty} (\frac{1}{R^2+z^2})^{h/2} dz \\
&= \Sigma(R) \\
&= \rho_0 \frac{\sqrt{\pi}}{R^{h-1}} \frac{\Gamma((h-1)/2)}{\Gamma(h/2)}.
\end{aligned}
\tag{4.13}
$$

where Gamma function is used and the integral can not be carried out to be an elementary function of the parameter h. The surface brightness at the galaxy center is $+\infty$ which is inconsistent with the Sérsic law.

4.4   Models of Galaxy Global Patterns: the Elliptical-shape Solutions

In this section I present rotationally symmetric and triaxial models of elliptical galaxy structures.

(i) **A model of 3-dimensional rotationally symmetric distributions.** The model is a corrected form of the above 3-dimensional light distribution (4.12),

$$
\begin{aligned}
f_{1s}(x,y,z) &= (h/2)\ln \frac{1}{(R_0+R)^2+(z_0+|z|)^2}, \\
\rho_{1s}(x,y,z) &= \rho_0 \left(\frac{1}{(R_0+R)^2+(z_0+|z|)^2}\right)^{h/2} \\
0 &< R < +\infty, \ -\infty < z < +\infty
\end{aligned}
\tag{4.14}
$$

where $R_0(>0)$ and $z_0(>0)$ are constants. This is equivalent to cutting out an infinite long cylinder of density distribution (of radius R_0) around the z-axis and cutting out an infinite plane layer of density distribution (of width $2z_0$) parallel to and centering on x-y plane from the spherically symmetric distribution (4.12). The corresponding orthogonal coordinate system is demonstrated in Fig. 4.2 which is the one in Fig. 4.1 with corresponding parts being cut out. The corresponding coordinate equation system is

$$
\begin{aligned}
&R = \lambda/(\lambda^2 + \mu^2) - R_0, \\
&|z| = |\mu|/(\lambda^2 + \mu^2) - z_0, \\
&0 < \lambda < R_0(R_0^2 + z_0^2)/z_0^2, \ 0 < |\mu| < (R_0^2 + z_0^2)/z_0.
\end{aligned} \tag{4.15}
$$

We can see that the 3-dimensional light distribution (4.14) is rotationally symmetric and its projection on any sky plane is apparently elliptical except on the x-y plane.

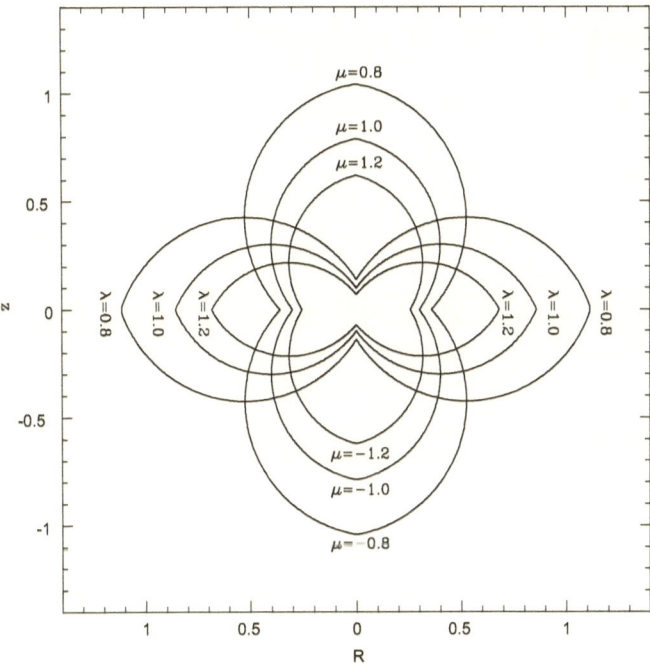

Figure 4.2: The orthogonal coordinate system is obtained by simply cutting out two infinite columns of areas in Fig. 4.1 centered on the $R = 0$ axis and the $z = 0$ axis respectively. The two columns are parallel to and centering on the Cartesian R-axis and z-axis respectively. The widths of the two columns are $2R_0$ and $2z_0$ respectively.

Now, we project the 3-dimensional light distribution on the sky plane which is a plane containing the x-axis and is obtained by rotating the x-y plane about the x-axis with an angle θ_0 (i. e., the inclination angle θ of the rotationally symmetric galaxy, see Fig. 4.3). Therefore, x'-axis coincides with x-axis and is a principal axis of inertia of the galaxy. The change of θ_0 includes all the possibilities between the galaxy seen head-on (circular-shaped) and edge-on (elliptical). The coordinates (x', y') on the sky plane (see Fig. 4.3) and the coordinate z' along the line-of-sight are related to the original coordinates (x, y, z) as follows

$$x = x',$$
$$y = y' \sec \theta_0 + z' \sin \theta_0, \qquad (4.16)$$
$$z = z' \cos \theta_0.$$

The surface brightness on the sky plane after the projection is given as the following integration,

$$\Sigma(x', y') = \int_{-\infty}^{+\infty} \rho(x, y, z) dz'$$
$$= \rho_0 \int_{-\infty}^{+\infty} \left(\frac{1}{(R_0 + \sqrt{x^2 + y^2})^2 + (z_0 + |z|)^2} \right)^{h/2} dz'. \qquad (4.17)$$

The x'-axis is the major or minor axis of the 2-dimensional elliptical light distributions $\Sigma(x', y')$. The integral can not be carried out to be an analytic formula. It can be numerically calculated. However, it can be analytically solved in some special cases.

Firstly, take $h = 3$, $\theta_0 = 0$, and $z_0 = 0$. The zero inclination $\theta_0 = 0$ implies that the symmetry axis is along the line of sight and the galaxy is seen head-on (circular-shaped galaxy). The corresponding surface brightness is

$$\Sigma(x', y') = 2\rho_0 / (R_0 + R')^2 \qquad (4.18)$$

where $R' = \sqrt{x'^2 + y'^2}$. This is exactly the Hubble law ([15]). This is not a coincidence. Our rational model of elliptical galaxies must be some truth!

Secondly, take $h = 3$, $\theta_0 = 0$, and $z_0 \neq 0$. The corresponding surface brightness is

$$\Sigma(x', y') = \frac{2\rho_0}{(R_0 + R')^2} \left(1 - \frac{z_0}{\sqrt{(z_0^2 + (R_0 + R')^2)}} \right) \qquad (4.19)$$

which is a generalized Hubble law. The general case $(h > 3)$ requires numerical calculation. However, the meaning of the parameters is simple. The model of rotationally symmetric density distributions (4.14) involves four parameters ρ_0, h, R_0, z_0. The central density of the distribution is proportional to ρ_0 while h represents galaxy density concentration and all other parameters $(R_0, z_0$ and x_0, y_0 in the following part (v)) affect the elliptical shapes of the 3-dimensional distributions. The projection of the distribution on to the sky plane (the integration (4.17)) can not be expressed by the Γ function of h as we did in (4.13). The total luminosity, i. e. the summation of the density distribution through whole space is finite when $h > 3$.

Fig. 4.3 is the cross section of the rotationally symmetric galaxy model (4.17) by the y-z plane which is also the y'-z' plane. x, y, z axes are the principle axes of inertia of the galaxy while x'-y' plane is the sky plane. The angle θ is the inclination angle θ_0 of the galaxy, and x-axis coincides with x'-axis. The change of θ_0 includes all the possibilities between the galaxy seen head-on (circular-shaped) and edge-on (elliptical). I choose the line-of-sight coordinate z' to be my integration variable in (4.21), which originates from y-axis, instead of the Cartesian coordinate $z'' = z' + y' \tan \theta_0$, which originates from y'-axis on the sky plane. This is because light density along line-of-sight takes its maximum at the galaxy principle axis: y-axis.

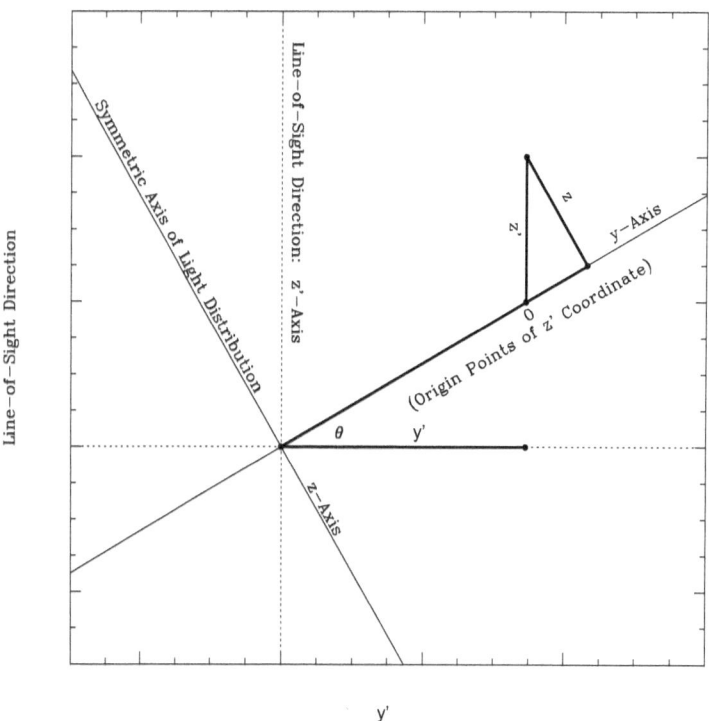

Figure 4.3: The cross section of the rotationally symmetric galaxy model (4.17) by the y-z plane which is also the y'-z' plane. x, y, z axes are the principle axes of inertia of the galaxy while x'-y' plane is the sky plane.

(ii) The 3-parameter rotationally symmetric model.

Our model of rotationally symmetric light distributions (4.14) involves 4 parameters (ρ_0, h, R_0, z_0). Its projection on to the sky plane (the integration (4.17)) can not be expressed by the Γ function of h as we did in (4.13). However, if we choose R_0 to be zero, our 4-parameter model changes into a 3-parameter model (ρ_0, h, z_0),

$$\rho_{2s}(x,y,z) = \rho_0 \left(\frac{1}{x^2 + y^2 + (z_0 + |z|)^2} \right)^{h/2} \tag{4.20}$$

and its projection on to the sky plane can be evaluated by the Γ function of h and the integrals on finite intervals,

$$\Sigma(x',y') = \rho_0 \int_{-\infty}^{+\infty} (\frac{1}{x^2+y^2+(z_0+|z|)^2})^{h/2} dz'$$
$$= \rho_0 (\int_0^{+\infty} (\frac{1}{x^2+y^2+(z_0+z)^2})^{h/2} dz'$$
$$+ \int_{-\infty}^{0} (\frac{1}{x^2+y^2+(z_0-z)^2})^{h/2} dz')$$
$$= \rho_0 \int_0^{+\infty} ((\frac{1}{R_+^2+(Z_++z')^2})^{h/2} + (\frac{1}{R_-^2+(Z_-+z')^2})^{h/2}) dz' \tag{4.21}$$
$$= \rho_0 (\frac{1}{R_+^{h-1}} \int_{Z_+/R_+}^{+\infty} (\frac{1}{1+s^2})^{h/2} ds$$
$$+ \frac{1}{R_-^{h-1}} \int_{Z_-/R_-}^{+\infty} (\frac{1}{1+s^2})^{h/2} ds)$$

where

$$R_\pm = \sqrt{x'^2 + (y' \mp z_0 \sin\theta_0)^2}$$
$$Z_\pm = \pm y' \tan\theta_0 + z_0 \cos\theta_0 \tag{4.22}$$

are not dependent on the integration variable z'. Now it is obvious why I choose the line-of-sight coordinate z' to be my integration variable, which originates from galaxy principle axis y-axis (see Fig. 4.3), instead of the Cartesian coordinate $z'' = z' + y' \tan\theta_0$, which originates from y'-axis on the sky plane. This is because light density along line-of-sight takes its maximum at the galaxy principle axis, y-axis, and the integration along line-of-sight (the formula (4.21)) breaks into two parts which have the formulations of the same type.

Finally, our 3 parameter model gives light distributions on

the sky plane as follows,

$$\Sigma(x', y') = \rho_0 \big($$
$$\frac{1}{R_+^{h-1}} \big(\frac{\sqrt{\pi}}{2} \frac{\Gamma((h-1)/2)}{\Gamma(h/2)} + \int_{Z_+/R_+}^{0} \big(\frac{1}{1+s^2} \big)^{h/2} ds \big)$$

$$+ \frac{1}{R_-^{h-1}} \big(\frac{\sqrt{\pi}}{2} \frac{\Gamma((h-1)/2)}{\Gamma(h/2)} + \int_{Z_-/R_-}^{0} \big(\frac{1}{1+s^2} \big)^{h/2} ds \big) \big). \tag{4.23}$$

This is exactly the model which is used in section 5 to be fitted to real elliptical galaxy images.

(iii) *Orthogonal coordinates and rational structure equation in the case of 3-dimensional space.* The curvilinear coordinate system in 3-dimensional space is

$$x = x(\lambda, \mu, \nu),$$
$$y = y(\lambda, \mu, \nu), \tag{4.24}$$
$$z = z(\lambda, \mu, \nu).$$

The arc-length derivatives are,

$$P = s'_\lambda = \sqrt{(x'^2_\lambda + y'^2_\lambda + z'^2_\lambda)},$$
$$Q = t'_\mu = \sqrt{(x'^2_\mu + y'^2_\mu + z'^2_\mu)}, \tag{4.25}$$
$$G = \sqrt{(x'^2_\nu + y'^2_\nu + z'^2_\nu)}.$$

We work on curvilinear coordinate system and the components of the gradient vector of the logarithmic density distribution $f(x, y, z)$ associated with the coordinate system are u, v and w. Because they are functions of the single variables λ, μ, ν respectively, the rational structure equation consists of the following equations,

$$u(\lambda) P'_\mu - v(\mu) Q'_\lambda = 0,$$
$$u(\lambda) P'_\nu - w(\nu) G'_\lambda = 0, \tag{4.26}$$
$$v(\mu) Q'_\nu - w(\nu) G'_\mu = 0.$$

Now we present the 6-sphere coordinates (λ, μ, ν) which is the generalization of the plane 4-circle coordinates (formulas (4.5) and Fig. 4.1) to the case of 3-dimensional space,

$$x = \lambda/(\lambda^2 + \mu^2 + \nu^2),$$
$$y = \mu/(\lambda^2 + \mu^2 + \nu^2),$$
$$z = \nu/(\lambda^2 + \mu^2 + \nu^2), \tag{4.27}$$
$$-\infty < \lambda, \mu, \nu < +\infty.$$

In Section 4.3, the plane 4-circle coordinate lines are rotated about the z-axis. The resulting 3-dimensional coordinate surfaces consist of one family of 2-spheres, one family of toroids, and the other family of all planes containing the z-axis. Here the coordinate system has triaxial symmetry and the coordinate surfaces consist of three families of 2-spheres. Therefore, the coordinate system is called a 6-sphere coordinate ([22]). The corresponding arc-length derivatives P, Q, G are

$$\begin{aligned} P(\lambda, \mu, \nu) &= 1/(\lambda^2 + \mu^2 + \nu^2) \\ &\equiv Q(\lambda, \mu, \nu) \equiv G(\lambda, \mu, \nu). \end{aligned} \tag{4.28}$$

The inverse formulas of the coordinate transformation are

$$\begin{aligned} \lambda &= x/(x^2 + y^2 + z^2), \\ \mu &= y/(x^2 + y^2 + z^2), \\ \nu &= z/(x^2 + y^2 + z^2), \\ &-\infty < x, y, z < +\infty. \end{aligned} \tag{4.29}$$

The rational structure equation (4.26) determines the directional derivatives (u, v, w) of our 3-dimensional logarithmic density distribution,

$$u(\lambda) = h\lambda, v(\mu) = h\mu, w(\nu) = h\nu \tag{4.30}$$

where h is a constant. The galaxy density distribution along all orthogonal coordinate lines can be found by performing path integrations of the directional derivatives,

$$\begin{aligned} f_n &= (h/2) \ln(\lambda^2 + \mu^2 + \nu^2) = (h/2) \ln \frac{1}{x^2+y^2+z^2}, \\ \rho_n(x, y, z) &= \rho_0(\frac{1}{x^2+y^2+z^2})^{h/2} = \rho_0 \frac{1}{r^h} \\ &-\infty < x, y, z < +\infty. \end{aligned} \tag{4.31}$$

We choose $h > 0$ because density $\rho = \rho_0 \exp f \to 0$ when $x, y, z \to +\infty$. The density distribution turns out to be the spherical-nucleus solution. However, we can achieve triaxial and other distributions based on the result, using the cut-out method.

(iv) **A model of spherically symmetric density distribution with zero gradient at the galaxy center.** We can cut out a sphere (radius a) of density distribution centered at the galaxy nucleus. The resulting distribution is

$$f_{3s}(x, y, z) = (h/2) \ln \frac{1}{(r+a)^2},$$
$$\rho_{3s}(x, y, z) = \rho_0 \frac{1}{(r+a)^h}, \tag{4.32}$$
$$0 < r < +\infty$$

In fact, it corresponds to the second factor of the Dehnen model (4.2) with $4 - \gamma$ being replaced by h.

(v) **A model of 3-dimensional triaxial density distributions.** We can cut out three infinite plane layers of density distributions parallel to and centering on the three Cartesian coordinate planes respectively from (4.31). The three planes are the x-y plane, y-z plane, and z-x plane. The widths of the three layers are $2x_0, 2y_0, 2z_0$ respectively. The resulting logarithmic density distribution is

$$f_{4s}(x, y, z) = (h/2) \ln \frac{1}{(x_0+|x|)^2+(y_0+|y|)^2+(z_0+|z|)^2},$$
$$\rho_{4s}(x, y, z) = \rho_0 \left(\frac{1}{(x_0+|x|)^2+(y_0+|y|)^2+(z_0+|z|)^2} \right)^{h/2}, \tag{4.33}$$
$$-\infty < x, y, z < +\infty$$

where $x_0(> 0), y_0(> 0)$ and $z_0(> 0)$ are constants.

Now, we project the 3-dimensional density distribution on the sky plane to see the surface brightness. Take the x-y plane to be the sky plane as an example. To achieve an analytic result, take $h = 3$. The resulting surface brightness is

$$\Sigma(x, y) = \frac{2\rho_0}{(x_0+|x|)^2+(y_0+|y|)^2} \left(1 - \frac{z_0}{\sqrt{(z_0^2+(x_0+|x|)^2+(y_0+|y|)^2)}} \right) \tag{4.34}$$

which is similar to the formula of the circularly symmetric light distribution (4.19). But here the surface brightness is non-circularly symmetric even when the sky plane is the x-y plane.

4.5 A Model of Full Galaxy Patterns

(i) **A Model of Full Galaxy Patterns.** To achieve a full account of the central areas as well as the global shapes, we need to add the logarithmic density distribution of the spherical-nucleus solution in section 4.3 to one of the logarithmic density distributions of elliptical-shape solutions in section 4.4. The summation of two logarithmic density distributions is equivalent to the multiplication of their direct density distributions, $\rho_n \rho_{is}$. If, instead of multiplication, we added together the direct density distributions, $\rho_n + \rho_{is}$, then we would see that elliptical galaxies consist of two apparent components. Real galaxy images, however, do not show such apparent components. The Dehnen model is recovered

$$\rho(r) = \rho_n(r)\rho_{3s}(r) = \rho_0 \frac{1}{r^{h_1}} \frac{1}{(r+a)^{h_2}} \qquad (4.35)$$

if we choose $h_1 = \gamma, h_2 = 4 - \gamma$. This gives a model of spherically symmetric patterns. The projected radial profile of the model density distribution resembles the de Vaucouleurs law. Its detailed discussion is given in [17]. All models $\rho_n \rho_{is}$, $i = 1, 2, 3, 4$, in the present Chapter have similar properties. Here, we pay special attention to the $\rho_n \rho_{2s}$ rotationally symmetric model

$$\begin{aligned} \rho(x, y, z) &= \rho_n \rho_{2s} \\ &= \rho_0 \frac{1}{r^{h_1}} \frac{1}{(x^2 + y^2 + (z_0 + |z|)^2)^{h_2/2}} \end{aligned} \qquad (4.36)$$

which gives account of both galaxy central area and elliptical shape. The model is called full 3-parameter rotationally symmetric model. The projection of the density distribution on to the sky plane can not be evaluated by the Γ function of h and the integrals on finite intervals as we did in (4.23) for the original 3-parameter model.

(ii) **Ellipticity as a function of radius.** Firstly, let us study the ellipticity of the projected isophote curves of the original 3-parameter model ρ_{2s}. The 3-dimensional density distribution of the model is the result of a spherically symmetric one with a layer of density centered on the x-y plane being cut out. Therefore, if the sky is a plane containing the z-axis then the projected light distribution (surface brightness) on the sky plane has approximately elliptical isophote curves (disky-shaped) whose ellipticity is approximately

$$\epsilon(R) = \sqrt{1 - R^2/(R + z_0)^2}. \qquad (4.37)$$

The ellipticity decreases monotonically with radius R from $\epsilon = 1$ at the galaxy center to $\epsilon = 0$ at $R = +\infty$. Actually, we see $1 > \epsilon_1 > \epsilon(R) > \epsilon_2 > 0$ because ρ_{2s} has a flat light distribution near the galaxy center and the image is always taken within a finite area.

Secondly, we study the full 3-parameter rotationally symmetric model, $\rho(x, y, z) = \rho_n \rho_{2s}$. The ellipticity $\epsilon(R)$ is a peaked function of R changing from $\epsilon = 0$ at the galaxy center to $\epsilon = 0$ at $R = +\infty$, because the spherical-nucleus solution has dominant logarithmic density ($\approx +\infty$) near the galaxy center while all elliptical-shape solutions have finite values (flat density distributions) near the galaxy center. Actually, we see an increasing ellipticity function if z_0 is large enough, or a decreasing function if z_0 is small enough, or a peaked or flat function if z_0 has intermediate values. These are consistent with Carter ([19])'s image analysis. The ellipticity function and the isophote-curve shapes (disky or boxy) change with the orientation of the line of sight and with the models of elliptical-shape solutions. This is not a coincidence. My rational model of elliptical galaxies must be some truth.

(iii) **Light-strength invariance.** Note that there is always an arbitrary factor ρ_0 accompanying our model light distribution (see, e. g. (4.13) and (4.14)). This is because our models involve the logarithmic density distributions $f(x, y, z)$ and the rational structure equation deals with their gradients only and leaves an integration constant for $f(x, y, z)$. This is called density-strength invariance which indicates that bright galaxies can share the same patterns with dim galaxies. Because galaxy visible mass distributions are approximately proportional to their light distributions, $\rho(x, y, z)$ can be mass distribution or light distribution in our model. In our fitting program, however, the projection of $\rho(x, y, z)$ on to the sky plane is fitted to the light distributions on galaxy images.

The satisfactory fitting of our model to real images (see section 4.6) suggests that elliptical galaxy patterns can be represented in terms of a few basic parameters given in the model. Real galaxies are further constrained by their physical process. For example, the shape of the radial profile and the shape of isophote curves change with luminosity of the objects. Our mathematical model gives larger freedom of degrees and bright galaxies are allowed to share the same patterns with dim galaxies.

4.6 Fitting the 3-Parameter Rotationally Symmetric Model to Elliptical Galaxy Images

Only the projections of the 3-parameter rotationally symmetric model ρ_{2s} and the triaxial model ρ_{4s} are found to be expressions of Γ functions and integrals on finite intervals. The projections of other models need further investigation. The direct numerical calculation of the integrals on infinite intervals takes significant computer time to obtain a reasonable result.

In the present section we fit the 3 parameter rotationally symmetric model (4.23) to real elliptical galaxy images and we find that the fitting is satisfactory except the areas near the galaxy centers. To achieve satisfactory fitting of full range of images, we need to apply the full model introduced in section 4.5 whose projection, however, is not an expression of Γ functions and integrals on finite intervals, and needs deep investigation.

The corresponding fitted elliptical pattern is shown in the middle columns of Figures 4.4 and 4.5). Note the two symmetric small spots on the central vertical direction of each fitted pattern in the Figures. This is due to almost zero values of R_+ and R_- at or near the two points (see (4.22)). Therefore, the integrals in (4.23) are on almost infinite intervals and their numerical calculations result in large errors by the computer which runs over a limited time. Now we know that it is important that we have an analytic model whose calculation can be taken on computers in a reasonable amount of time.

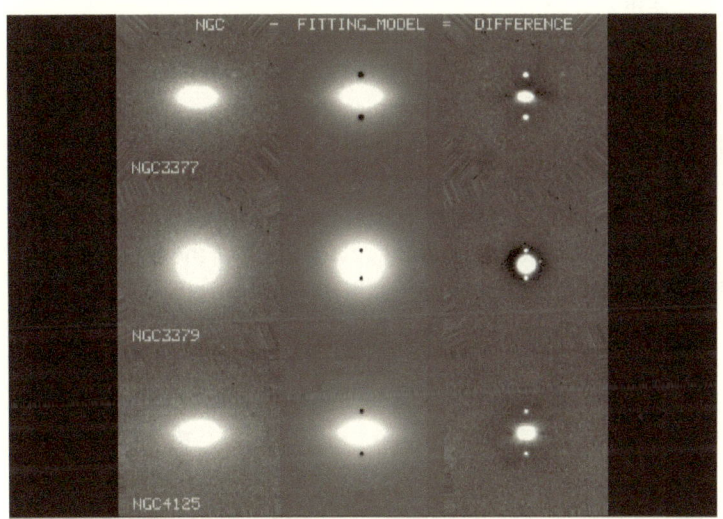

Figure 4.4: The Frei *et al.* R-band images [21], NGC 3377, 3379, 4125, minus our theoretical fitting patterns respectively result in the residual images. Explanation of the spots on the fitting images are given in the text.

Figures 4.4 and 4.5 demonstrate the 2-dimensional fitting result. We can see that the outer parts of images are fitted satisfactorily and the central parts are not. There are two major reasons. The first reason is that I should have used the full models, $\rho_n \rho_{is}$, introduced in section 4.5 which generalize Dehnen law (4.2). However, the projections of full models do not have analytic expressions or integrals on finite intervals. Their numerical fitting is left for able people who can devise efficient algorithms and utilize supercomputers. The fitting error in the third column of Figures 4 and 5 should be circularly-symmetric about galaxy centers so that they can be represented by the spherical-nucleus solution. However, the fitting errors from NGC4365 and NGC4406 are far different from circularly-symmetric patterns. The fitting errors of other galaxies are not perfectly symmetric too. These are due to the second reason that both the pixel light values and the signal-to-noise ratios are very large near galaxy centers, and the numerical calculation of the integrals in (4.23) by common computers result in large errors. Highly efficient algorithms and utilization of supercomputers can resolve the problem and eliminate the two symmetric small spots on the central vertical direction of each fitted pattern in Figures 4.4 and 4.5.

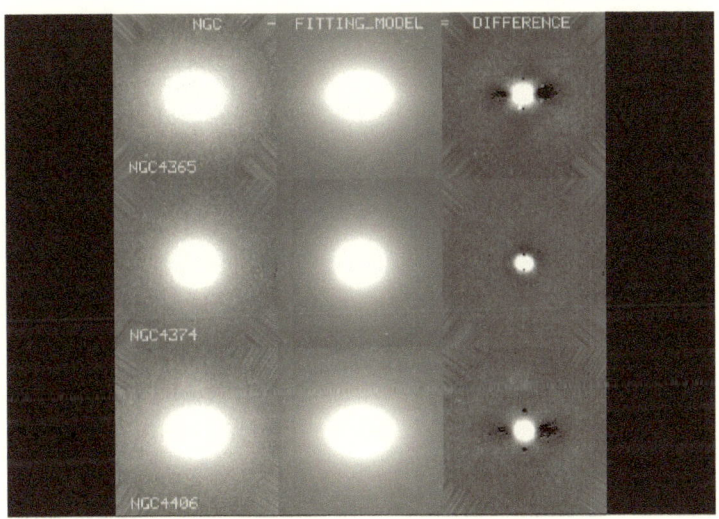

Figure 4.5: The Frei *et al.* R-band images [21], NGC 4365, 4374, 4406, minus our theoretical fitting patterns respectively result in the residual images. Explanation of the spots on the fitting images are given in the text.

65

Chapter 5

Why is there No Dust in Elliptical Galaxies?

5.1 Prop. 2: Triaxial Surfaces Always Lead to Spherically Symmetric Structures

As astronomers know, any component of any galaxy pattern (except the arm pattern) is either circularly symmetric, or bilaterally symmetric and the components are either two dimensional structures of exponential disks or galactic bars or three dimensional elliptical patterns. Amazingly, there are very few patterns of rational structures. What I have found is circularly symmetric or spherically symmetric. The only pattern I have found which is not circularly and spherically symmetric is the dual handle structure. This is not a coincidence. However, there are many types of orthogonal curves, and each complex analytic function corresponds to a net of orthogonal curves. Prop. 1 in Chapter 3 which needs to be proved, says that any net of orthogonal curves is either circularly symmetric with respect to the center point, or bilaterally symmetric. However, a net of orthogonal curves generally does not lead to a rational structure because the corresponding rational structure equation has no solution.

That is, the equation generally does not factor out two directional derivatives, $u(\lambda)$ and $v(\mu)$ respectively.

The requirement of three dimensional rational structure is even tighter because the rationality involves the proportion surfaces not the proportion lines in two-dimensional plane. For the purpose of heuristic educating, however, I used proportion lines to introduce the study on three dimensional rational structure. But only proportion surfaces can be defined in three dimensional rational structure.

The orthogonal proportion surfaces of triaxial symmetry (e.g., the 6-sphere coordinate system (4.27)) is a symmetry of higher degree than the rotational symmetry. Therefore, real elliptical galaxies may be triaxial. However, my mathematical sense leads to the following proposition: triaxial and smooth shapes of orthogonal proportion surfaces always lead to the rational structures of spherical symmetry. That is, the rational structure turns out to be spherically symmetric although its orthogonal proportion surfaces look triaxial. This is my proposition, which needs mathematical justification. If the proposition is true then the only way to get triaxial rational structure is to use non-smooth shapes of orthogonal proportion surfaces as we did with cutting-out method (see (4.33)).

5.2 Prop. 3: Orthogonal Proportion Surfaces of Triaxial Symmetry have no Other Proportion Surface

Assume a three dimensional rational structure is derived from an orthogonal net of proportion surfaces. My proposition is that the structure has no other proportion surface which makes an acute angle to the original surfaces.

In the case of an orthogonal net of proportion lines on a plane, however, the proposition is not true. For example, the rational structure of exponential disks has the proportion line of any equiangular spiral of any angle. The second example, the rational structure of dual handles, has any proportion line of any spiral provided that the spiral makes a fixed angle to the confocal ellipses and hyperbolas but the spirals may not form a net of orthogonal curves.

If the above proposition is proved then it may explain why there is no dust in elliptical galaxies. The proportion surfaces of elliptical galaxies are the intersecting nets of orthogonal spheres and there is no other proportion surface. Accordingly, disturbing waves are difficult to form and spread. On the other hand, spiral galaxies are two-dimensional and their proportion curves are open spirals where disturbance waves are easy to form and spread. Astronomical observations do show that arms do not exist in elliptical galaxies.

The disturbance to rational structure leads to the formation of gas and dust. New families of stars and planets are born to these gas and dust. The star-planet families are short-lived. This happens only in spiral galaxies.

5.3   Prop. 4: Elliptical Galaxies are Completely Rational Structure

Although exponential disk and dual handle structure are each the rational structure, their density addition, $\rho_d + \rho_b$, may not be a rational structure. That is, the addition of rational structures may not be rational too. Similarly, we add the logarithmic density distribution of the spherical-nucleus solution in section 4.3 to one of the logarithmic density distributions of elliptical-shape solutions in section 4.4 to achieve a full account of the elliptical galaxy density profile as well as the global shapes: $f_n + f_{is}$. This model of elliptical galaxies might not be a rational structure either. But rational structure is based on the directional derivatives of logarithmic density and the addition of two logarithmic density distributions is equivalent to the multiplication of their direct density distributions, $\rho_n \rho_{is}$. The multiplication of rational structures may be rational too in the case of our model of elliptical galaxies. However, this is just a mathematical proposition. Please help us prove that elliptical galaxies are completely rational structure.

Appendix A

Equiangular Coordinate System as a Uniquely Defined Mapping between Coordinate Spaces.

We did not specify the variance domain S on (λ, μ) coordinate plane on which the equiangular coordinate system (2.14) is defined and maps it onto the whole (x, y) coordinate plane. To find the domain, we define two constants which are called periods,

$$
\begin{aligned}
\Delta_\lambda &= \frac{2\pi d_2}{d_1 d_4 - d_3 d_2} \ (> 0), \\
\Delta_\mu &= \frac{2\pi d_1}{d_1 d_4 - d_3 d_2} \ (> 0).
\end{aligned}
\tag{A.1}
$$

It is straightforward to show that the two periods satisfy the following equations,

$$
\begin{aligned}
d_1 \Delta_\lambda - d_2 \Delta_\mu &= 0, \\
-d_3 \Delta_\lambda + d_4 \Delta_\mu &= 2\pi.
\end{aligned}
\tag{A.2}
$$

There are many such domains and now we want to prove the statement of coordinate periodicity that any vertical infinite band S_{λ_1} on (λ, μ) plane can be mapped onto the whole (x, y) plane of a galaxy disk,

$$S_{\lambda_1} : \lambda_1 < \lambda < \lambda_2 (= \lambda_1 + \Delta_\lambda), \quad -\infty < \mu < +\infty. \quad (\text{A.3})$$

where λ_1 is arbitrary constant and the length of the interval (λ_1, λ_2) is the period,

$$\Delta_\lambda = \lambda_2 - \lambda_1. \quad (\text{A.4})$$

As indicated in Fig. 2.1, it is equivalent to showing that the two different vertical boundary lines, $\lambda = \lambda_1$ and $\lambda = \lambda_2$ on (λ, μ) plane are mapped to a single curvilinear coordinate line on (x, y) plane (see Fig. 2.1). This can be shown by three steps. We choose a closed curve (thick dotted line in Fig. 2.1) which consists of two sections, one from the coordinate lines of first set ($\mu = $ constant $= \mu_1, \lambda_1 < \lambda < \lambda_2$) and the other from the second set ($\lambda = $ constant $= \lambda_1, \mu_1 < \mu < \mu_2$). The closed curve is called snail-shaped curve (see Fig. 2.1).

For the first step, we show that the arc-length derivative $Q(\lambda, \mu)$ is uniquely defined along the snail-shaped curve. Starting at the point N_1 where the value of Q is $Q(\lambda_1, \mu_1)$, we have two ways to go to the point N_2, one being the section Γ and the other Υ. The corresponding two values of Q at N_2 are $Q(\lambda_1, \mu_2)$ and $Q(\lambda_2, \mu_1)$ respectively. The uniqueness of Q means that $Q(\lambda_1, \mu_2) = Q(\lambda_2, \mu_1)$. This is guaranteed by the first equation in (A.2).

For the second step, the values of the tangent angle $\alpha(\lambda, \mu)$ of the coordinate line $\lambda = \lambda_1$ (see Fig. 2.1) should be uniquely defined along the line within a difference of 2π. Starting at the point N_1 where the value of α is $\alpha(\lambda_1, \mu_1)$, we have two ways to go to the point N_2, one being the section Γ and the other Υ. The corresponding two values of α at N_2 are $\alpha(\lambda_1, \mu_2)$ and $\alpha(\lambda_2, \mu_1)$ respectively. Therefore,

$$(\alpha(\lambda_1, \mu_2) - \alpha(\lambda_1, \mu_1)) - (\alpha(\lambda_2, \mu_1) - \alpha(\lambda_1, \mu_1)) = 2\pi. \quad (\text{A.5})$$

Now we need calculate the value of $\alpha(\lambda, \mu)$. Its calculation is straightforward with formulas (2.14),

$$\tan\alpha = \frac{y'_\mu}{x'_\mu} = \frac{d_2\sin\theta + d_4\cos\theta}{d_2\cos\theta - d_4\sin\theta}. \tag{A.6}$$

Defining

$$\cos\theta_0 = \frac{d_2}{\sqrt{d_2^2 + d_4^2}}, \quad \sin\theta_0 = \frac{d_4}{\sqrt{d_2^2 + d_4^2}}, \tag{A.7}$$

then we have

$$\tan\alpha = \tan(\theta + \theta_0). \tag{A.8}$$

Finally we find the value of the tangent angle of the coordinate line $\lambda =$ constant

$$\alpha = \theta + \theta_0 = d_3\lambda + d_4\mu + \theta_0. \tag{A.9}$$

Substituting the value into the equation (A.5), we find that the equation reduces to the second equation in (A.2) which is already proved. Note that, in Fig. 2.1, α is the tangent angle of the coordinate line $\lambda =$ constant at a point (x, y) and θ is the polar angle of the same point. Their difference is the pitch angle of the coordinate line $\lambda =$ constant. Therefore, we find the pitch angle i to be,

$$i = \alpha - \theta = \theta_0 \tag{A.10}$$

which is constant over the whole plane. We have proved that our coordinate lines (proportion curves) are equiangular spirals.

For the third step, we need prove that the snail-shaped curve is really closed. To see the case, we change the coordinate transformation equations (2.14) into different form, making use of the formulas (2.17) and (A.9),

$$x = Q(\lambda, \mu)\cos(\alpha - \theta_0)/\sqrt{d_2^2 + d_4^2},$$
$$y = Q(\lambda, \mu)\sin(\alpha - \theta_0)/\sqrt{d_2^2 + d_4^2}. \tag{A.11}$$

Because $Q(\lambda, \mu)$ and α are uniquely defined and both θ_0 and $\sqrt{(d_2^2 + d_4^2)}$ are constants, x, y are uniquely defined. That is, the snail-shaped curve is really closed.

Therefore, we have shown that the two different vertical boundary lines, $\lambda = \lambda_1$ and $\lambda = \lambda_2$, on (λ, μ) plane are mapped to a single curvilinear coordinate line on (x, y) plane (see Fig. 2.1). This is equivalent to say that the vertical infinite band S_{λ_1} on (λ, μ) plane (see (A.3)) is mapped onto the whole (x, y) plane of the galaxy disk. Because λ_1 is arbitrarily chosen, we have shown that the coordinate transformation (2.14) is periodic in that each interval of λ of length (period) Δ_λ with $-\infty < \mu < +\infty$ is mapped onto whole (x, y) plane. The periodicity of μ can be similarly proved.

Bibliography

[1] http://web.ipac.caltech.edu/
staff/jarrett/galaxies/spirals.html

[2] http://www.tng.iac.es/
news/2000/07/06/m51/

[3] He, J. & Yang, X. 2009, The Origin Of Natural Structure
(Indiana: AuthorHouse)

[4] He, J. 2003, Astrophys. & Space Sci., **283**, 305

[5] He, J. 2004, Astrophys. & Space Sci., **291**, 163

[6] He, J. & Yang, X. 2006, Astrophys. & Space Sci., 302, 7

[7] He, J. 2006, Astrophys. & Space Sci., **305**, 197

[8] He, J. 2008, Astrophys. & Space Sci., **313**, 373

[9] He, J. 2005, http://www.arxiv.org/
abs/astro-ph/0510535

[10] He, J. 2005, http://www.arxiv.org/
abs/astro-ph/0510536

[11] He, J. 2005, http://adsabs.harvard.edu/
abs/2005PhDT........17H

[12] Martinez-Valpuesta, I., Knapen, J. H. & R. Buta, R.
2007, Astron. J., **134**, 1863

[13] Eskridge, et al. 2002, Astrophys. J. S., **143**, 73

[14] Sérsic, J.: 1968, *Atlas de Galaxias Australes. Córdoba. Obs. Astronómico*

[15] Hubble, E. P.: 1930, *ApJ*, **71**, 231

[16] Young, P. J.: 1976, *AJ*, **81**, 807

[17] Dehnen, W.: 1993, *MNRAS*, **265**, 250

[18] Stark, A. A.: 1977, *AJ*, **213**, 368

[19] Carter, D.: 1978, *MNRAS*, **182**, 797

[20] Binney, J. and Merrifield, M.: 1998, *Galactic Astronomy*, Princeton Univ. Press, Princeton, NJ

[21] Frei, Z., Guhathakurta, P., Gunn, J. E., and Tyson, J. A.: 1996, *AJ*, **111**, 174

[22] Moon, P. and Spencer, D. E.: 1961, *Field Theory Handbook*, Springer-Verlag, Berlin